YOUR KNOWLEDGE HAS VALUE

- We will publish your bachelor's and
  master's thesis, essays and papers

- Your own eBook and book -
  sold worldwide in all relevant shops

- Earn money with each sale

Upload your text at www.GRIN.com
and publish for free

Bibliographic information published by the German National Library:

The German National Library lists this publication in the National Bibliography; detailed bibliographic data are available on the Internet at http://dnb.dnb.de .

Imprint:

Copyright © 2015 GRIN Verlag, Open Publishing GmbH
Print and binding: Books on Demand GmbH, Norderstedt Germany
ISBN: 9783668256101

This book at GRIN:

http://www.grin.com/en/e-book/335244/3d-segmentation-and-boundary-completion-using-subjective-surface-method

Lina Gundelwein

Aus der Reihe: e-fellows.net stipendiaten-wissen

e-fellows.net (Hrsg.)

Band 1980

3D segmentation and boundary completion using subjective surface method

GRIN Publishing

Friedrich-Alexander-University
Erlangen-Nürnberg

Technical Faculty

3D segmentation and boundary completion using subjective surface method

Computational Engineering - TAF Optics

**for the academical degree
Bachelor of Science**

Topic: 3D segmentation and boundary completion using subjective surface method

Author: Lina Gundelwein

Version of: March 2, 2015

Contents

List of Figures

List of Tables

List of Algorithms

1. Intro

Segmentation is used to locate boundaries of an object in a given image. Finding those boundaries is important e.g. for visualizing three dimensional objects and measuring the surface or the volume of an object. Ideally, the boundaries coincide with the edges seen in the image. However, due to noise, partly missing information or in case of subjective contours the edges can be irregular and discontinuous. The human brain is able to perform visual completion that is to say it can in many cases complete those interrupted boundaries by filling in the missing gaps. Simple segmentation methods however fail to do so ([MS07], [SMS00]). Based on the level set method [Set99] an intrinsic model of deformable surfaces was proposed by Vicent Caselles et al. [CCCD93] and by Ravi Malladi et al. [MSV95] where the surface propagates by an implicit velocity depending on the image gradient and therewith stops at the boundary. This allowed the detection of interior and exterior contours of objects with boundaries given by gradients. Further research then lead to the subjective surface method that is presented in [SMS00] and that allows for the detection of missing boundaries in 2D. It sharpens the surface around the edges and connects segmented boundaries across the gaps. All those methods use finite differences to solve the resulting initial value partial differential equation. As opposed to this Michael Fried is using finite elements with the subjective surface method resulting in an algorithm that detects interrupted boundaries as well as subjective contours in 2D [FM09]. His algorithm is based on Karol Mikula's idea of using the subjective surface method for medical image segmentation [MS07].

As an extension of Michael Fried's work, the algorithm presented in this thesis should be able to complete missing boundary segments in 3D, smooth out the contours and be able to perform modal completion.

The remainder of this paper is divided into three sections: at first we give the necessary mathematical background comprising the level set method and finite element method, see I. In Part II, we present the implementation of the problem with the finite element toolbox ALBERTA. Finally, in Part III we present the experimental results for the segmentation of synthetic and real 3D images, elaborating on noisy images and modal completion. The paper ends with a short concluding section.

Part I.

Mathematics

The program extended in the scope of this works implements the subjective surface method solved with finite elements. The subjective surface method for 3D image segmentation presented in [MS07] is related to geodesic mean curvature flow of level sets. In the following, first the level set method and its extension to the subjective surface method are described. Then the discretization of the model and the solution via finite elements are presented.

2. Level Set Method

A simple approach for image segmentation is to take a segmentation function and let it evolve over time so it assumes the shape of the segmented object. We start with a small initial curve or surface, e.g. a sphere, and let it grow until it locks on the boundary of the object we want to segment. This way we can isolate an object from its background.

The level set method devised in 1987 by Stanley Osher and James A. Sethian describes the moving curves or surfaces [OF01], [Set99]. The method computes and tracks the motion of an evolving curve or interface Γ in two or three dimensions, see Figure 1. The interface Γ separates two regions from each other and moves with the scalar speed v in normal direction n. The evolution over time of any point x on such a level set curve therefor is characterized by $x'(t) = vn$.

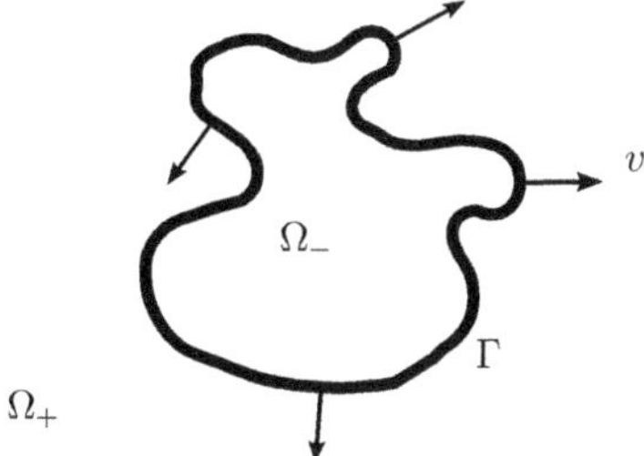

Figure 1: Level set curve with normals

The speed function v can depend on local geometric information such as curvature and normal direction, shape and position of the front, time and other shape independent properties like external physics. Is $v > 0$, the interface moves forward that is in normal direction and for $v < 0$ it moves backward, in opposite normal direction. The initial position of the front is the zero level set of the smooth level set function u. This function's evolution can be linked with the propagation of the interface itself. The zero level set of u has to match the propagating front at any time. This implies

$$u(x(t), t) = 0.$$

To derive an equation of the motion we use the chain rule

$$u_t - \nabla u(x(t), t) \cdot x'(t) = 0,$$

with $x'(t) = vn$, where $n = \frac{\nabla u}{|\nabla u|}$ this results in the following evolution equation

$$u_t - v|\nabla u| = 0. \tag{1}$$

In sum, we want to solve

$$u_t - v|\nabla u| = 0$$
$$\Gamma(t) = \{(x, y, z) | u(x, y, z, t) = 0\}$$

in an image domain $\Omega \in \mathbb{R}^3$ with zero Neumann boundary condition at its boundary surface $\partial\Omega$.

2.1. Speed function

We choose the speed function

$$v = \kappa = \nabla \cdot \frac{\nabla u}{|\nabla u|},$$

where κ is the mean curvature of the surface. This means in regions of high curvature the front will move fast, whilst in regions of small curvature the interface moves slow or stops. [Set99]

2.2. Edge detector

Gage and Grayson showed that a simple closed curve moving under its curvature is shrinking to a point and in three dimensions a hyper-surface moving under its mean curvature analogously shrinks to one or more spheres until it disappears [Set99]. As we do not want the curves to vanish, but halt at the object boundaries we need our speed function to be zero at the boundaries. For this purpose we will multiply the speed function by an edge function. The edge function detects the boundaries by taking the gradient of the intensity function I_0 and in that way determines when the surface evolution needs to stop. We use the edge detector

$$g = \frac{1}{1 + |\nabla I_0|},$$

where I_0 denotes the original image [Set99]. It is strictly positive in homogeneous regions and close to zero on the edges. The curve is attracted by these detected edges and a steady state of it is taken as the boundary of the segmented object. However, if the contours are interrupted the algorithm fails as the curve only stops at ideal edges where $g = 0$.

Therefor, Caselles and Kimmel introduced a new gradient term

$$\nabla g = \nabla \left(\frac{1}{1 + |\nabla I_0|} \right)$$

in [CKS97] that increases the attraction of the curve or surface towards the boundary and allows to track boundaries with a high variation in their gradient and small gaps.

2.3. Regularization

In noisy images, problems still occur because the advection is insufficient and the noise components may act like edges, attracting the evolving curve or surface. This can be avoided by adding a regularization term. Thus, in order to work with noisy images, we add a curvature dependence acting as a regularization to the normal velocity v to achieve sufficient advection and to smooth out noise and inhomogeneities. We are using the regularization term $g\kappa$. The geometrical equation for v after multiplying the speed function with the edge detector and adding the regularization term is given by

$$v_{reg} = g\kappa + \nabla g \cdot n.$$

Plugging the regularized normal velocity v_{reg} into (1) results in the level set formulation of the surface evolution [Fri03], [MS07]:

$$u_t - g|\nabla u|\nabla \cdot \frac{\nabla u}{|\nabla u|} + \nabla g \cdot \nabla u = u_t - |\nabla u|\nabla \cdot \left(g\frac{\nabla u}{|\nabla u|} \right) = 0. \tag{2}$$

This equation describes an active contour model with which one can detect objects with interior holes. It allows for cusps, corners and automatic topological changes. Using this approach on the zero level set, the initial surface has to be close to the final shape to give reasonable results, otherwise the surface might not be driven there. As a consequence we do not only consider the evolution of the zero level set, but the whole hyper surface u that is composed by all level sets, leading to the subjective surface method [SMS00], [SMS02]. With the evolution over time all level sets will move towards the boundaries and build up there.

In this section we show how to approximate equation 2 with finite elements. The finite element method uses a simple approximation of unknown variables to transform partial differential equations into algebraic equations.

2.4. Finite element space

Let $\mathbb{T}_h$ be a conforming simplicial triangulation of $\Omega \in \mathbb{R}^n$ and $\mathbb{P}_1(S)$ the space of linear polynomials on the simplex S, then we define the finite element space X_h by

$$X_h := \{\varphi \in C^0(\bar{\Omega}) | \varphi \in \mathbb{P}_1(S) \forall S \in \mathbb{T}_h\}. \tag{3}$$

We use the barycentric coordinates

$$\lambda(x) = (\lambda_0, \ldots, \lambda_d)(x) \in \mathbb{R}^{d+1}$$

to describe finite elements on the simplicial grid. The barycentric coordinates of $x \in \mathbb{R}^d$ can be obtained by solving the SLE

$$\sum_{j=0}^{d} \lambda_j a_j = x, \quad \sum_{j=0}^{d} \lambda_j = 1.$$

The reference simplex is defined by

$$\bar{S} := \{(\lambda_0, \ldots, \lambda_d) \in \mathbb{R}^{d+1} | \lambda_k \geq 0, \sum_{k=0}^{d} \lambda_k = 1\},$$

which equals the standard simplex $\hat{S}$, if $a_0 = 0, a_j = e_j, j = 1, \ldots, d$.

Each simplex $S \subset \mathbb{T}_h$ can be mapped to the standard simplex $\hat{S}$ by the parameterization

$$F_S : \hat{S} \to S, F_S(\hat{x}) = A_S \hat{x} + a_0, \tag{4}$$

where the matrix $A_S \in \mathbb{R}^{n \times d}$ is given by

$$A_S = \begin{pmatrix} \vdots & & \vdots \\ a_1 - a_0 & \cdots & a_d - a_0 \\ \vdots & & \vdots \end{pmatrix}.$$

Now the non vanishing global basis functions are locally numbered on the simplex S. The local basis functions $\{\bar{\varphi}^1, \ldots, \bar{\varphi}^m\}$ are coupled with the global basis functions

$\{\psi_1, \ldots, \psi_N\}$, where $N = dim X_h$. For every basis function ψ_j that does not vanish on $S \in \mathbb{T}_h$ holds

$$\psi_{j|S}(x(\lambda)) = \bar{\varphi}^i(\lambda),$$

where $i \in \{1, \ldots, m\}$ depends on j and S. m equals the number of local basis functions on an element. Following the notation given in the ALBERTA documentation [SS05], we now denote $i_S : J_S \to \{1, \ldots, m\}$ to be the mapping of the global index set J_S to the index set of the local basis functions on S. With $j_S : \{1, \ldots, m\} \to J_S$ being the inverse mapping of i_S, the connection between local and global numbering on each element S is determined by

$$\psi_j(x(\lambda)) = \bar{\varphi}^{i_s(j)}(\lambda) \; \forall \lambda \in \bar{S}, \, j \in J_S$$

$$\psi_{j_S(i)}(x(\lambda)) = \bar{\varphi}^i(\lambda) \; \forall \lambda \in \bar{S}, \, i \in \{1, \ldots, m\}.$$

There is a local coefficient vector $(u_S^1, \ldots, u_S^m) = (u_{j_S(1)}, \ldots, u_{j_S(m)})$ of u_h on S next to the global coefficient vector $(u_1, \ldots, u_N)$.

The local representation of u_h on the simplex S can thus be written as

$$u_h(x) = \sum_{i=1}^{m} u_S^i \bar{\varphi}^i(x) \; \forall x \in S$$

or as the finite element functions are usually not evaluated at world coordinates x but at barycentric coordinates λ on S:

$$u_h(x(\lambda)) = \sum_{i=1}^{m} u_S^i \bar{\varphi}^i(\lambda(x)) \; \forall x \in \bar{S}. \tag{5}$$

2.5. Discretization

2.5.1. Regularization

Before starting the discretization we slightly change the diffusion term in the segmentation model. To avoid problems with a vanishing gradient ∇u, we use the Evans-Spruck regularization [ES$^+$91] substituting $|\nabla u|$ by $\sqrt{\varepsilon^2 + |\nabla u|^2}$, with $\varepsilon \in (0, 1)$ being a positive regularization parameter:

$$u_t - \sqrt{\varepsilon^2 + |\nabla u|^2} \nabla \cdot \left(g \frac{\nabla u}{\sqrt{\varepsilon^2 + |\nabla u|^2}} \right) = 0. \tag{6}$$

This regularization ansatz has been successfully used for finite element algorithms dealing with level set formulations of mean curvature flow related problems, c.f.[DD01], [DD03],[Fri04]. The parameter ε determines whether mean curvature flow of level sets ($\varepsilon = 0$) or mean curvature flow of the graphs ($\varepsilon = 1$) is performed. For segmentation of objects with irregular and interrupted boundaries ε should be chosen to be very close to zero. Large variations in the graph of the segmentation function, that is noise components in the input image, are smoothed due to large mean curvature [MS07].

2.5.2. Weak formulation

We now reformulate the problem (6) into the weak form which is a variational statement of the problem. We multiply with a test function $\varphi \in H^1(\Omega)$, integrate over the the domain Ω and then use integration by parts using zero Neumann boundary conditions. This relaxes the problem. Instead of finding a pointwise solution, we find a solution that satisfies the strong form on an average over the domain.

$$\int_\Omega \frac{u_t}{\sqrt{\varepsilon^2 + |\nabla u|^2}}\varphi + \int_\Omega g\frac{\nabla u}{\sqrt{\varepsilon^2 + |\nabla u|^2}} \cdot \nabla\varphi = 0. \tag{7}$$

2.5.3. Spatial discretization

Based on the weak formulation we want to solve the problem locally on elements. Using the finite element space (3), the spatial discretization of the weak identity (7) is given by

$$\int_\Omega \frac{u_{t,h}}{\sqrt{\varepsilon^2 + |\nabla u_h|^2}}\varphi_h + \int_\Omega g\frac{\nabla u_h}{\sqrt{\varepsilon^2 + |\nabla u_h|^2}} \cdot \nabla\varphi_h = 0.$$

Note that in our case $u_h = u_h(t, x)$ is not only location-dependent but also time-dependent. In the finite element discretization the time dependence can be found in the coefficients:

$$u_h(t, x(\lambda)) = \sum_{i=1}^{m} u_S^i(t)\bar\varphi^i(\lambda(x)) \, \forall x \in \bar{S}.$$

2.5.4. Time discretization

The backward Euler scheme is used for time discretization. The time step size is denoted by $\tau = \frac{T}{M}$. With $m = 0, 1, \ldots, M$ time steps, $\Psi^m = \Psi(t^m)$ defines any function $\Psi : \Omega \times [0, T] \to \mathbb{R}$ at the time $t^m = m\tau$. Taking the nonlinear terms of the equation from the previous time step and considering the linear terms on the current time level, we get a semi-implicit time discretization [MS07] of the weak identity (7)

$$\frac{1}{\tau} \int_\Omega \frac{u_h^m - u_h^{m-1}}{\sqrt{\varepsilon^2 + |\nabla u_h^{m-1}|^2}}\varphi_h + \int_\Omega g\frac{\nabla u_h^m}{\sqrt{\varepsilon^2 + |\nabla u_h^{m-1}|^2}} \cdot \nabla\varphi_h = 0$$

for $m = 1, 2, ..., M$, where u_h^0 is a suitable finite element approximation to the given initial value $u_h(0)$. For better readability, the subscript h is dropped from now on writing u^m and φ instead of u_h^m and φ_h.

We now split the first order term and bring u^{m-1} from the old time step to the right hand side:

$$\frac{1}{\tau}\int_\Omega \frac{u^m}{\sqrt{\varepsilon^2+|\nabla u^{m-1}|^2}}\varphi + \int_\Omega g\frac{\nabla u^m}{\sqrt{\varepsilon^2+|\nabla u^{m-1}|^2}}\cdot\nabla\varphi = \frac{1}{\tau}\int_\Omega \frac{u^{m-1}}{\sqrt{\varepsilon^2+|\nabla u^{m-1}|^2}}\varphi. \tag{8}$$

Using local representation and the abbreviations

$$c = \frac{1}{\sqrt{\varepsilon^2+|\nabla u^{m-1}|^2}}, \quad A = \frac{g}{\sqrt{\varepsilon^2+|\nabla u^{m-1}|^2}}$$

we can rewrite (8) as the LSE

$$\frac{1}{\tau}\sum_{j=1}^{N} u_j^m \int_S c\varphi_j\varphi_i + \sum_{j=1}^{N} u_j^m \int_S \nabla\varphi_j A\cdot\nabla\varphi_i = \frac{1}{\tau}\sum_{j=1}^{N} u_j^{m-1}\int_S c\varphi_j\varphi_i,$$

where i depends on S and j.

By applying the index mapping $i_S : J_S \to \{1,\ldots,m\}$ and transformation (4) to the standard simplex we can rewrite the equation again to

$$\frac{1}{\tau}\sum_{j=1}^{N} u_j^m \int_{\hat S} \bar c\bar\varphi^i(\lambda(\hat x))\bar\varphi^j(\lambda(\hat x)) + \sum_{j=1}^{N} u_j^m \int_{\hat S} \nabla_\lambda\bar\varphi^j(\lambda(\hat x))\cdot \bar A\lambda(\hat x))\nabla_\lambda\bar\varphi^i(\lambda(\hat x) =$$

$$\frac{1}{\tau}\sum_{j=1}^{N} u_j^{m-1}\int_{\hat S} \bar c\bar\varphi^i(\lambda(\hat x))\bar\varphi^j(\lambda(\hat x)), \tag{9}$$

using the abbreviations

$$\bar c = \frac{|\det DF_S|}{\sqrt{\varepsilon^2+|\nabla u^{m-1}|^2}} = |\det DF_S|c, \tag{10}$$

where DF_S is the Jacobian of F_S, and

$$\bar A = |\det DF_S|\Lambda(\hat x(\lambda))\frac{g}{\sqrt{\varepsilon^2+|\nabla u^{m-1}|^2}}\Lambda^T(\hat x(\lambda)) = |\det DF_S|\Lambda(\hat x(\lambda))A\Lambda^T(\hat x(\lambda)), \tag{11}$$

where $\Lambda = \Lambda_S \in \mathbb{R}^{d\times n}$ is the Jacobian of the barycentric coordinates on S:

$$\Lambda(x) := \begin{pmatrix} \lambda_{0,x_1}(x) & \lambda_{0,x_2}(x) & \ldots & \lambda_{0,x_n}(x) \\ \vdots & \vdots & & \vdots \\ \lambda_{d,x_1}(x) & \lambda_{d,x_2}(x) & \ldots & \lambda_{d,x_n}(x) \end{pmatrix} = \begin{pmatrix} \nabla\lambda_0(x)^T \\ \vdots \\ \nabla\lambda_d(x)^T \end{pmatrix}.$$

We calculate the integrals by quadrature. A numeric quadrature is a set of pairs of weights and points with

$$\int_{\hat{S}} f(\hat{x})d\hat{x} = \hat{Q}(f) := \sum_{k=0}^{n_Q-1} \omega_k f(\hat{x}(\lambda_k)), \tag{12}$$

where n_Q is the number of quadrature points. Using the affine transformation described in (4), we can get a quadrature for any simplex S by transforming (12):

$$\int_{S} f(x)dx = Q_S(f) := \hat{Q}((f \circ F_S)|\det DF_S|) = \sum_{k=0}^{n_Q-1} \omega_k f(x(\lambda_k))|\det DF_S(\hat{x}(\lambda_k))|.$$

Replacing the integrals in (9) by the quadrature terms

$$M = \frac{1}{\tau} \sum_{k=0}^{n_Q-1} \omega_k \left(\bar{c}\left(\lambda\left(\hat{x}\right)\right) \bar{\varphi}^i\left(\lambda\left(\hat{x}\right)\right) \bar{\varphi}^j\left(\lambda\left(\hat{x}\right)\right) \right),$$

$$C = \sum_{k=0}^{n_Q-1} \omega_k \left(\left(\nabla_\lambda\bar{\varphi}^j\left(\lambda\left(\hat{x}\right)\right) \cdot \bar{A}\left(\lambda\left(\hat{x}\right)\right) \nabla_\lambda\bar{\varphi}^i\left(\lambda\left(\hat{x}\right)\right)\right) \right)$$

yields

$$[M + C] \sum_{j=1}^{N} u_j^m = M \sum_{j=1}^{N} u_j^{m-1}. \tag{13}$$

M and C will from now on be called mass and stiffness operator respectively.

Part II.

Implementation

In this part the implementation of the time dependent problem and solution of the discrete system (13) as well as the input and output methods are described.

Starting with the initialization of all parameters, structures and functions that are needed for the further calculations, the finite element space is set up and the volume to be segmented is read in. After initial mesh adaption the solution of the time dependent problem starts. In every time step first the image I_0 is interpolated on the mesh and then the adaptive algorithm is run, adapting the mesh appropriately and solving the discrete problem on it. At the end of every time step η_τ which is the L2-norm of the difference between the old and the current time step is calculated and the finite element function is regularized.

Those steps are repeated until η_τ comes below a given tolerance δ. Then the finite element method is evaluated and results are put out.

The program structure is shown in Algorithm 1.

Algorithm 1 Program structure

1: Initialization of parameters, structures and functions $\rightarrow$ section 4
2: Adapt the initial mesh
3: **while** $\eta_\tau > \delta$ **do**
4:     Set $u_{n-1} = u_n$, interpolate I_0 on X_h
5:     Do one time step $\rightarrow$ algorithm 2
6:     Calculate $\eta_\tau = ||u_{n-1} - u_n||_2$, regularize u_n
7:     $n = n + 1$
8: **end while**
9: Write u_{n-1} to output files $\rightarrow$ section 6

Adaptive algorithm $\rightarrow$ section 5

Below the basic concepts of the toolbox we are using are presented followed by the detailed description of the algorithm.

3. Basic Concepts of ALBERTA

The finite element algorithm is realized with the finite element toolbox ALBERTA. In this part the toolbox is introduced and the main concepts are presented. ALBERTA is an adaptive multilevel finite element toolbox developed in ANSI-C. The library provides data structures and functions for adaptive finite element simulations in 1-D, 2-D and 3-D. ALBERTA can be used for fast and flexible implementation of efficient software for real life applications and simulations. It is based on modern algorithms like adaptive methods, higher order discretizations, fast linear and non-linear iterative solvers and multi-level algorithms. Furthermore you can improve numerical methods with ALBTERTA and directly integrate them in existing simulation software [SS05].

3.1. Mesh

The computational domain is triangulated into tetrahedra. A tetrahedra mesh as opposed to a rectangular one simplifies the adaptive mesh refinement that is going to be used. It is easier to subdivide elements without causing hanging nodes. Furthermore they are more versatile as they are body-fit to moving interfaces. We start with the macro triangulation which equals the coarsest mesh. To build the simplicial grid it can then either be refined at the beginning or it can be adapted dynamically with the evolving solution. The adapted parts can be coarsened again, provided that the macro triangulation represents the coarsest possible grid. ALBERTA uses the Kossaczký refinement algorithm which leads to nested meshes with a hierarchical structure of binary trees [SS05]: A refined element is refined into two children elements. Therefor every element is the root of a binary tree where the leaf nodes represent the current triangulation. If elements are marked to be coarsened, they are simply coarsened back into their parent element.

3.2. Degrees of freedom

Each leaf element stores information about its local degrees of freedom (DOFs) that connect finite element data with geometric information of the triangulation. A DOF is realized as a simple integer index. Thus when refining the mesh and adding new DOFs the index range has to be enlarged, too. During mesh coarsening DOFs are removed and thereby holes in the list of used indexes might appear. All vectors or matrices storing data on the DOFs have to be adjusted in their size as well. ALBERTA provides a DOF administration tool that automatically re-sizes matrices and vectors accordingly and ensures a continuous indexing by compressing the index range, that is renumbering the DOFs.

3.3. Adaptive mesh refinement

The aim of the adaptive mesh refinement is that the used mesh is adapted to the problem in order to fulfill a given criterion like an error tolerance *tol* which is compared to the global error estimate which is given by the sum of the local error estimates:

$$||u - u_h|| \leq \eta(u_h) = \left(\sum_{S \in \mathbb{T}} \eta_S(u_h)^p \right)^{1/p}, \; p \in [1, \infty).$$

This equation holds true if a efficient and reliable error estimator exists. In our case we do not have an error estimator but an error indicator, so it is not guaranteed that the refining really will decrease the global error. However this does not pose a problem as our aim is not to find the perfect solution of the level set function u_h, but to have a high resolution in the vicinity of the boundaries.

The finer the mesh is, the higher the resolution will be. On the other hand a finer mesh requires more memory space and causes a higher computational effort and thereby higher computation times. Therefore the mesh should be as fine as necessary and as coarse as possible.

Refining the whole mesh leads to the best error reduction, but might not be necessary. Some parts with fine structures might need a high mesh resolution whilst for regions of homogeneous intensity it is not necessary to refine the mesh. Global refinement would produce more unknowns than needed to reduce the error below a given tolerance. To save computation time and memory space we only want to refine the regions requiring more resolution due to high curvature or rapidly changing speed functions and coarsen parts that do not need high resolution. This process is called local refinement.

At the same time some parts of the mesh might be finer than necessary and can therefor be coarsened. This will lead to a larger error, but it will also decrease the amount of unknowns and therefor speed up the further computations. The marking strategy checks how much the error increases when coarsening the elements and if it is less than a given tolerance,it will be coarsened. In Figure 2 it shows the evolution of the number of DOFs with and without coarsening for the artificial test data set "Test Spheres" by Stefan Röttger[Röt06]. There are different strategies implemented in

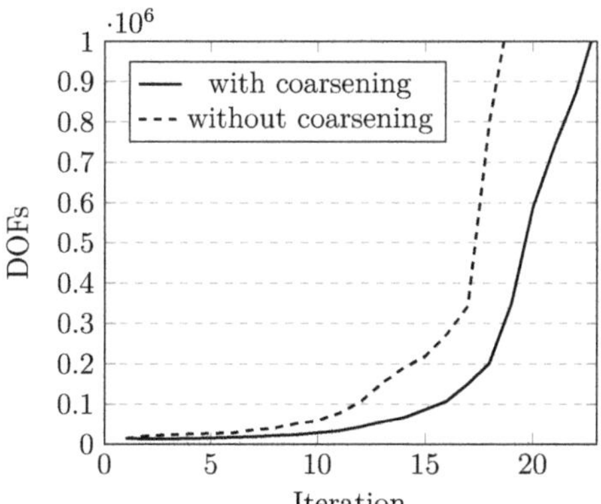

Figure 2: Number of DOFs

ALBERTA for deciding which elements have to be refined and which need coarsening.

4. Initialization of parameters and structures

At the beginning all parameters have to be initialized. Some of the parameters are pre-set in the code others change from problem to problem so they are read in from a separate initialization file. Furthermore the 3D image data has to be read in.

4.1. Parameters

There are some parameters that we want to be able to tweak without having to compile the code again. Those are e.g. the parameters defining name of the input, output and macro files and the number of smoothing steps, but also the parameters that are used by the ALBERTA routines for the solution of the LSE and for mesh adaption. Those are pre-initialized with default values, but we would like to initialize some of them differently. The parameters that should differ from the default values as well as the other parameters we might want to tweak are included in a parameter file that is read in at the beginning. You can find an example file in appendix IV.

4.2. Finite element space

ALBERTA offers Lagrange finite elements up to order four. We are using piecewise linear finite elements which are defined by their values at the vertices of the triangulation, so each (tetrahedron) element is defined by four basis functions. The basis functions and DOFs together with the underlying mesh define the finite element space. The underlying mesh holds all the information about the triangulation. The data for the macro triangulation is read in from a simple ASCII-file that gives information about the number of vertices and elements, the element boundaries and neighbors and determines the refinement edges. We use a macro file defining a standard triangulation of a cube in $\mathbb{R}^3$ centered at the origin with edge length 2. The cube has eight vertices and six elements, all meeting at one diagonal that is marked as the refinement edge. In Figure 3 this would be the line connecting node 0 and node 5.

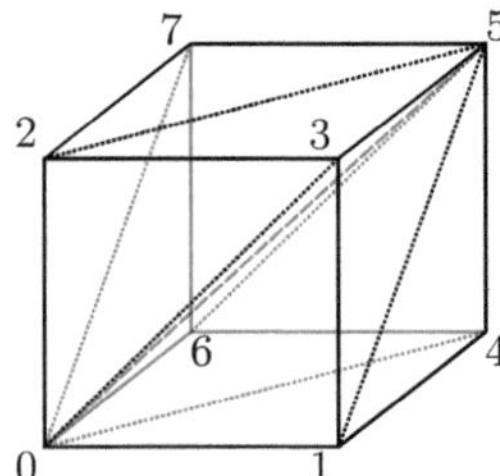

Figure 3: Macro triangulation

4.3. Initial data of the equation

The boundary values and right hand side function f are set to zero. The simulation is started with an initial function u_0. Assuming that the segmented object is placed in a central position, the initial function is given as a peak in the center of the domain. Different functions can be used, all of them having their peak with value 1 in the center point and decreasing with growing distance from the center down to value 0. For most of our tests we set the initial function to $u_0 = 1 - |x|^2$ which can be seen in Figure 4.

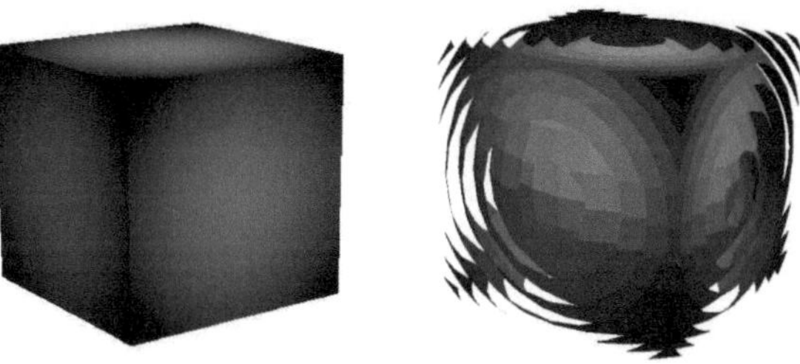

Figure 4: The initial level set function u_0 (left) and only ten level sets of the initial function (right).

If smaller objects are to be segmented the initial data function should be adjusted, see 7.4.

4.4. Input image

We work with gray scale images representing slices of the volume we want to segment. Those images are read in and their intensity value are stored and mapped on to finite element space. In the plain PGM format the pixels' intensity values are given in decimal notation formatted as ASCII characters. The file starts with the header giving the maximum gray value *maxval* and the *height* and *width* of the image. It is followed by a raster of *height* rows consisting of *width* gray values. Each gray value is a number from zero through *maxval*, where zero equals black and *maxval* equals white. The *depth* of the volume is given by the number of PGMs and is read in from the parameter file.

The gray scale images in plain PGM format can be defined as intensity maps that assign an intensity $I(x_1, x_2)$ to each pixel (x_1, x_2). Composing the intensity maps of the 2D slices, leads to an intensity map $I_{3D}(x_1, x_2, x_3)$ of the volume.

It is assembled by looping over all PGMs and saving its gray values in a 3D array that is dynamically allocated. For each PGM first the header is read to obtain the *height* and *width* and then the gray values are saved by looping through the rows. In the process the overall maximum gray value $MAXVAL$ is detected and saved.

The intensity values have to be mapped onto the discrete mesh which underlies the finite element space to obtain the intensity map $I(x, y, z) = I_0$ that assigns an intensity value to every DOF. If the used triangulation is not of a unit cube the offset has to be considered. Assuming that d_x, d_y, d_z are the diameters of the mesh in the respective dimensions, the mapping assigns a normalized intensity value $\in (0, 1)$ to every point (x, y, z) on the discrete mesh:

$$I(x, y, z) = \frac{I_{3D}(x_1, x_2, x_3)}{MAXVAL},$$

where

$$x_1 = \frac{width}{d_x}(x - x_{offset}), \quad x_2 = \frac{height}{d_y}(y - y_{offset}), \quad x_3 = \frac{depth}{d_z}(z - z_{offset}).$$

5. Adaptive algorithm

After reading in all input parameters and data the actual calculations can begin.

5.1. Initial mesh adaption

Before starting the adaptive algorithm, the initial mesh has to be adapted to the problem. That is to say the mesh is adjusted to fit the given volume that we want to segment. The mesh should be fine in the neighborhood of edges and coarse in continuous regions. As not all edges are visible in the image itself the mesh will be further adjusted later on, but at least where the edges are visible the mesh should be appropriately refined. For initial adaption the equidistribution strategy (ES) has turned out to give good results. In some cases, especially if the image is not very detailed, the guaranteed error reduction strategy (GERS) gives a sufficiently fine resolved grid with a smaller amount of DOFs, which speeds up the simulation. In Figure 5 you can see a slice of a simple and a complex 3D image mapped on the initial mesh before refinement and after refinement with either ES or GERS strategy.

5.2. Mesh adaption

As the problem to be solved is time dependent, the mesh is adapted in every time step using a posteriori error estimators. We use an explicit strategy where the time step is constant and the method to mark which elements have to be refined or coarsened is GERS. In each time step the mesh is adapted based on the error estimate computed from the discrete solution u_h from the previous time step. The DOF indexes are compressed automatically after mesh adaption. Then the current time step is solved on the adapted mesh (see section 5.3) and the error estimators are computed by a call to an ALBERTA intern function.

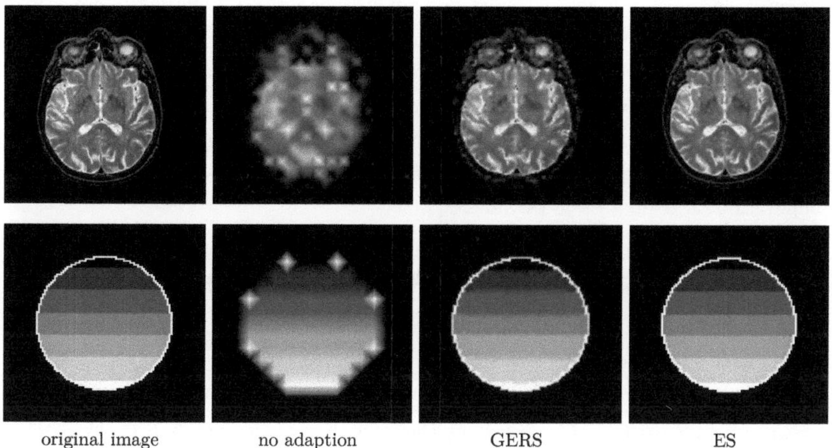

Figure 5: One slice of a complex (top) and a simple (bottom) image interpolated on the mesh after initial adaption with different adaption strategies.

This leads to Algorithm 2 for one time step.

Algorithm 2 One time step

1: Start with given tolerance *tol*, time step size τ, the triangulation $T_{t,n}$ from the previous time step t and the error estimate η computed from the discrete solution $u_{t,n}$ on $T_{t,n}$
2: $T_{t+\tau,0} = T_{t,n}$
3: $n := 0$
4: $t := t + \tau$
5: **while** $\eta > tol$ **do**
6:     Mark elements for refinement or coarsening using GERS
7:     Adapt mesh $T_{t,n}$ producing $T_{t,n+1}$
8:     Build LSE: Assemble system matrix and right hand side $\rightarrow$ algorithm 3
9:     Solve the discrete problem for $u_{t,n+1}$ on $T_{t,n+1}$
10:    Compute error estimate η
11:    $n := n + 1$
12: **end while**

After each time step the finite element function is normalized so that the values of the finite element functions lie between 0 and 1 and the L2-norm $||u_{n-1} - u_n||_2 = \eta_\tau$ of the difference between the old and the current time step is calculated. So while η is the tolerance for the local mesh adaption that is done in each time step, η_τ acts as the stopping criterion for the whole algorithm, see Algorithm 1.

As the algorithm gets slower and consumes a lot more memory space with an increasing amount of unknowns, it might be necessary to restrict the amount of DOFs by disabling the refinement if the amount of DOFs exceeds a given limit *maxDOFs*.

The mesh can now still be coarsened but not further refined. Only if the amount of DOFs falls below $maxDOFs$ again, the refinement is re-enabled. At a certain point no more coarsening is possible and the amount of DOFs is above $maxDOFs$, so the algorithm will keep working with the DOFs given at that point. This will result in lower quality and resolution of the solution than a calculation with an unlimited amount of unknowns, but it speeds up the computation on small data sets and it makes the computation on big data sets possible.

5.3. Assemblage and solution of the discrete system

As established in Part I the problem is solved by quadrature on finite elements, leading to the linear system of equations (13) built up by the stiffness and mass operators that are defined on each element. We want to solve the discrete system, where the system matrix $L = C + M$ as well as the right hand side vector $f = M \sum_{j=1}^{N} u_j^{m-1}$ need information from the mass and/or stiffness matrix. Those matrices are updated in every step by looping over all elements, calculating the local contribution and adding it to the global matrices. The local contribution of each element is calculated by calling functions that evaluate $\bar{A}$ (11) and $\bar{c}$ (10) at the given quadrature points. This includes the evaluation on all quadrature nodes of the edge detector g as well as the square root term that is calculated with the discrete solution u^{m-1}.

Algorithm 3 Assemblage

1: Initialize functions to calculate $\bar{A}$ and $\bar{c}$
2: **for all** Elements $S \in \mathbb{T}$ **do**
3:     Calculate $\bar{A}_S$ and $\bar{c}_S$
4:     Calculate integrals C_S and M_S
5:     Add C_S and M_S to global matrix L and M_S to global matrix f
6: **end for**

For the solution of the linear system ALBERTA offers an orthogonal error method (OEM) interface. The LSE $Lu^m = f$ (13) is solved with a Conjugate Gradient solver.

6. Output

To display a final level set u^m, we use different methods.

For once ALBERTA provides a routine that converts the ALBERTA mesh structure into an unstructured tetrahedron grid saved in VTK format that can be displayed with ParaView. ParaView is an open-source application that facilitates interactive 3D data visualization. We mainly use the clip tool, contours filter and threshold filter for the extraction of sub-regions and isosurfaces and the histogram filter to display the distribution of the intensity values. To compare consecutive time steps a programmable python filter can be applied.

The other tool we are using is the volume renderer V^3: Versatile Volume Viewer that uses the pre-integration technique to display volume data. The volumes color, opacity and intensity can be adjusted interactively by drawing transfer functions. The histogram of the volume is depicted by default. The tool is suited for highlighting different parts of the model in different colors to distinguish between the real boundaries and subjective surfaces that have been detected with the algorithm and for displaying several semi-transparent layers at the same time.

V^3 offers a routine to convert a set of PGMs to a pvm file which will then be displayed by the volume renderer. Thus only the PGM output had to be programmed. It is basically the reversion of the input algorithm. By default PGMs in the same size ($width *$ $height$), quantity ($depth$) and with the same range of gray values ($0, \ldots MAXVAL$) as the ones read in the input routine are given out. A 3D array is dynamically allocated by looping over $depth, height$ and $width$. The evaluation of the finite element functions is done locally on single elements using barycentric coordinates. Hence to read out the value at a certain world coordinate, the element (simplex) at that location and the corresponding barycentric coordinates have to be found. This is done by calling an ALBERTA intern function with the respective coordinate (x, y, z), where

$$x = x_{offset} + (k + 0.5) * \frac{d_x}{width}, \ k \in \{0, 1, \ldots width\}$$

$$y = y_{offset} + (j + 0.5) * \frac{d_y}{height}, \ j \in \{0, 1, \ldots height\}$$

$$z = z_{offset} + (i + 0.5) * \frac{d_z}{depth} \ i \in \{0, 1, \ldots depth\}.$$

After obtaining the element and the barycentric coordinates of the given point, the value of the finite element function at this location is calculated by calling another ALBERTA function and it is saved into the array at the position (i, j, k). The 3D array can be seen as an array of 2D arrays. Each 2D array is written into one PGM file, resulting in $depth$ PGMs.

While ParaView offers many tools to visualize and analyze the data it runs very slow on big data sets, especially when it comes to volume visualization and transparency. V^3 on the other hand provides smooth volume presentation, but less tools for quantitative analysis of the intensity values.

Part III.
Results

In this section we present the results we obtained using our model on several synthetic and real images, aimed at performing modal completion on objects with interrupted boundaries. The tests are run on a 64bit operating system Ubuntu in Version 12.04 carried out on an 8 core processor of type Intel(R) Xeon(R) CPU E5-2680 0 @ 2.70GHz with 20480kB (20MB) Intel® Smart Cache and 16347480 kB (16GB) main memory thereof 12814408 kB (12,22GB) available.

7. Synthetic 3D images

Testing on synthetic images has the advantage that image defects like missing pixels, holes in the boundaries, noise or blur can be added purposely and in a controlled way. The results can be easily evaluated as the correct solution is known beforehand. All synthetic images used here are created programmatically by the author.

7.1. Boundary completion

The aim was to fill in missing boundaries and perform modal completion. The algorithm will find the existent boundaries as well as the modal boundaries which might be perceived by the human visual system although in fact they are not visible in the image. The Kanizsa triangle, see Figure 6, is a classical example for modal completion.

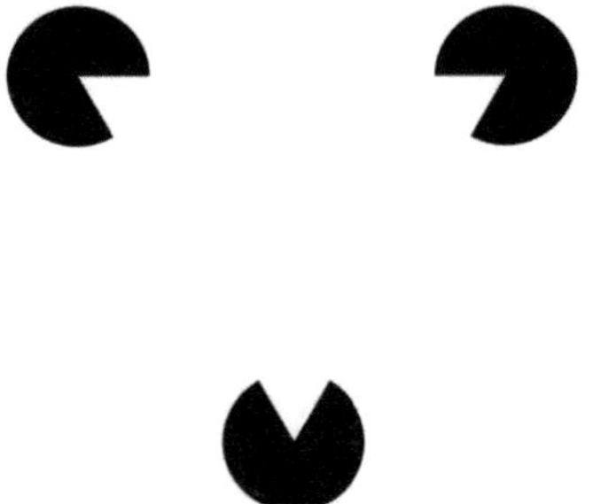

Figure 6: The Kanizsa triangle

It shows that the human brain can complete objects with boundaries that are not characterized by gradients: One can perceive a triangle with well defined contours although only three circles with a missing sixth are depicted.

The algorithm by Michael Fried was able to detect those subjective contours in the Kanizsa triangle [FM09] . Now this example can be transformed into a 3D problem - a "Kanizsa pyramid" or alternatively a "Kanizsa cube". The Kanizsa pyramid consists of 4 spheres each missing a pyramid shaped part and the Kanizsa cube consists of 8 spheres each missing an eighth.

The algorithm was run with the number of DOFs restricted to 200.000. The pyramid and cube are correctly segmented as to be seen in Figure 7.

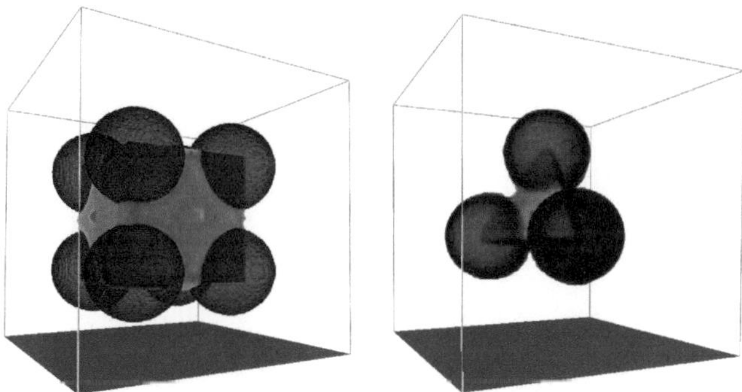

Figure 7: Segmented Kanizsa cube and pyramid segmented to precision $\delta = 0.001$.

A problem that occurs is that the grid is very fine where the edges are visible in the initial image, but too coarse in the areas of modal completion, see Figure 8. This is the case because in the initial adaption the mesh is adapted to be very fine where the real boundaries in the image were. The number of DOFs then soon exceed $maxDOFs$ so the areas of modal boundaries are not refined. Running the algorithm with a higher $maxDOFs$ will give more accurate results.

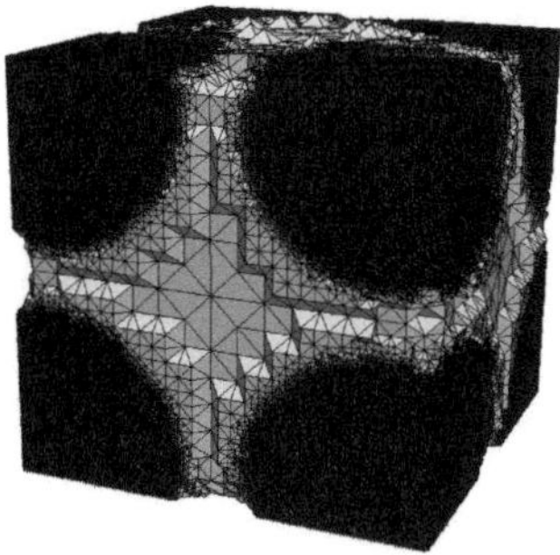

Figure 8: Grid on the segmented Kanizsa cube after removing the spheres via threshold filter in ParaView.

The results also showed, that a too coarse resolution of the input images has a negative influence on the quality of the modal completion. The boundaries will not be smooth and the pixel edge length is too long in comparison to the image itself. In such a case the modal boundaries will not be detected if they do not coincide with the pixel edges.

Another example is shown in Figure 9. The image is composed of two gray levels, showing a cube with a hole in its surface. The algorithm detects the existing boundaries as well as the missing ones. In the resulting image the intensity will converge to three

Figure 9: The complete segmentation of a cube with a quadratic hole on one side (left) and extracted with a threshold filter: the segmented real boundary (middle) and the inside of the cube, giving the inner boundary (right).

different values: the background intensity converges to zero, the values on the existing boundary of the cube approximates 0.75 and the inside of the cube has an intensity approximating 1 as can be seen in Figure 10. This means all level sets with a value between 0 and 0.75 built up at the outer boundary of the cube and all level sets with a value greater than 0.75 meet at the inner and the modal boundary of the cube.

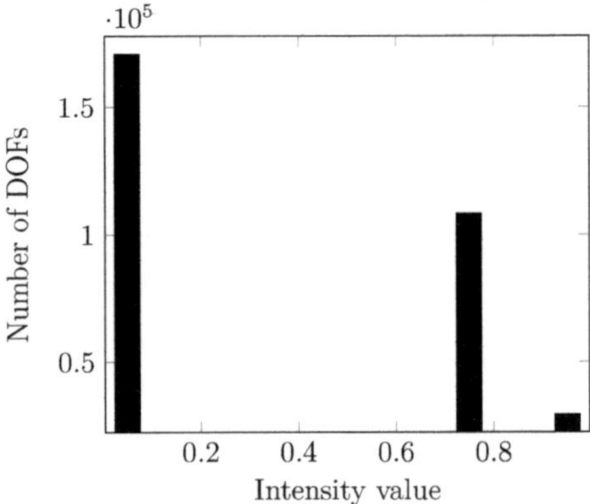

Figure 10: Histogram showing the amount of DOFs with a certain intensity value.

7.2. Nested objects

Although the algorithm is not constructed to find boundaries of several separate objects, they can be found if the boundaries of the two objects are not too close to each other. Also, boundaries of nested objects can be detected. Figure 11 shows a clip of the detected boundaries of a hollow cube with a massive sphere inside. The background intensity converged to 0. All level sets with a value between 0 and 0.88 are built up at the outer boundary of the cube. On the inner boundary of the cube the level sets with an intensity greater than 0.88 and smaller or equal to 0.93 meet. All level sets with a higher value than 0.93 define the boundary of the sphere.

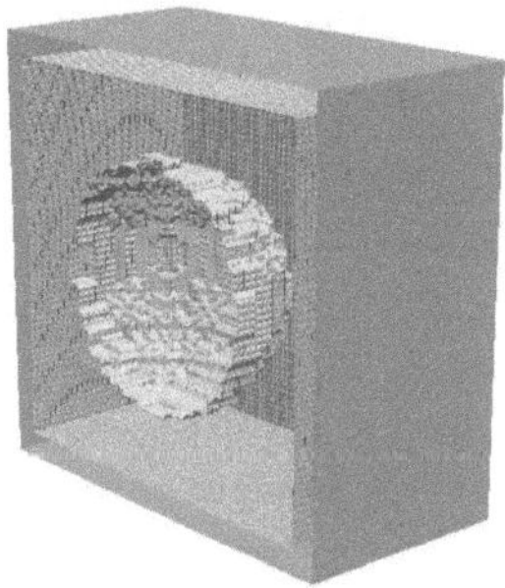

Figure 11: Clip of the segmented boundaries of a massive sphere inside a hollow cube.

7.3. Noisy or blurred images

Many images contain blurry areas, noise and other artifacts due to how they were acquired.

As shown in Figure 12, the algorithm can detect blurry edges, but the boundaries of the segmented object will not look as smooth as they would if the input image was clear.

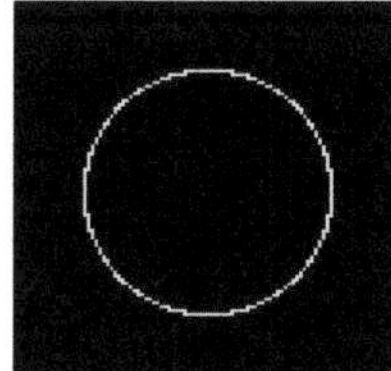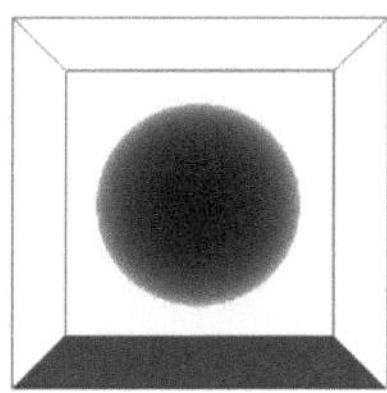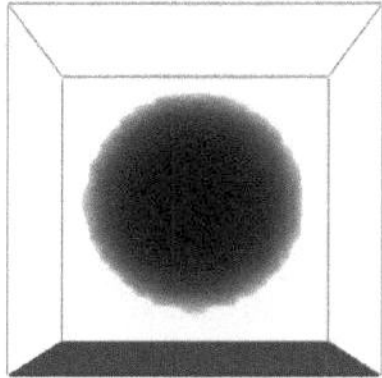

Figure 12: Slice of the input volume and segmentation of a clear (left) and blurred sphere (right).

How the algorithm handles the different kinds of noise is tested on volumes that are deliberately corrupted with noise. First we look at random noise which is characterized by fluctuations of the intensity above and below the image intensity. If the noise is low pronounced, that is the noise pixels' intensity only slightly varies from the correct intensity, the algorithm does not have problems finding the correct edges and can be run as if no noise existed, see Figure 13.

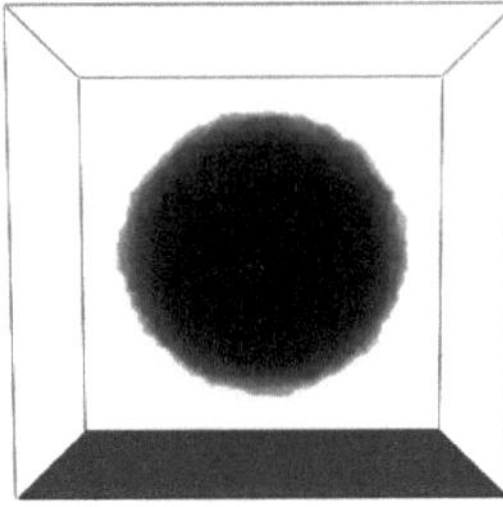

Figure 13: Slice of a blurred sphere with low pronounced random noise (left) and the segmented object (right).

If the intensity fluctuations are very high so the noise pixels' intensity varies a lot from the correct intensity, the algorithm detects them as boundaries and tries to segment them. To avoid that those noise pixels are treated the same as real boundaries they need to be smoothed out at the beginning. This can easily be done by setting the edge detector $g = 1$ for the first few steps. This way the level sets move under their mean curvature and do not stop on the noise boundaries. However, if the number of smoothing steps is too large, the real boundaries will vanish and into the bargain the number of needed iterations increases with the number of smoothing steps. Consequently the number of smoothing steps should be as low as possible.

The smoothing is tested on an image of a white sphere on black background located in the center of the volume, see Figure 14. A number of 3 smoothing steps turned out to be sufficient to remove all noise and segment the sphere in 33 iterations. The same sphere without smoothing needs 30 iterations to be segmented, so the smoothing did not affect the computation time much, it just removed the noise. Running the same case with 10 smoothing steps needs 49 iterations, without giving a better result than the 3-steps-smoothing. With noise, the number of DOFs after initial adaption is already very high, see Table 1. This results in very high computation times and is nearly impossible to calculate on one processor due to the big amount of unknowns that need a lot of memory space.

If the random noise is of a very low frequency while having high intensity fluctuations it cannot be removed with the same technique, because too many smoothing steps would be needed. To fully remove it the smoothing would remove the real boundaries, too. So while the algorithm can remove high frequency random noise, images with low

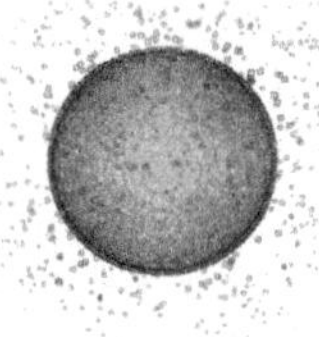  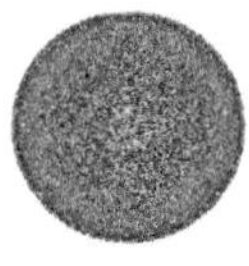

Original Image   0 smoothing steps,   3 smoothing steps,   10 smoothing steps,
                  30 iterations   33 iterations   49 iterations

Figure 14: 0.1% of the pixels are set to maximum intensity.

Noise	DOFs
0.0%	2.354.424
0.1%	2.376.589
0.2%	3.173.238
0.5%	4.490.991

Table 1: Number of DOFs after initial adaption for different noise components in a 100x100x100 px volume of a sphere with radius 33 px.

frequency noise have to be preprocessed. The same applies to other artifacts like ring or metal artifacts, beam hardening and scatter. For further information on artifacts see [BF12], [BK04].

7.4. Small objects

If the objects to be segmented are very small in comparison to the domain it is recommended to use an initial function that decreases faster and/or is masked as it will reduce the number of needed iterations.

Here, we will just look at the masking. Assuming the domain is quadratic, if we want to segment a cube with an edge length of $width/4$ that is located in the center of the domain described in Section 3, we know that there are no boundaries to be detected outside of a radius $r = \sqrt{2 \cdot (\frac{width}{2 \cdot 4})^2}$. Hence we can set the initial function to

$$u_0 = \begin{cases} 1 - |x|^2 & \text{if } r < k \\ 0 & \text{if } r \geq k. \end{cases}$$

Good results are achieved for $k \in (r, width/2)$, the best for $k = width/4 = 0.5$, see Figure 15. Without masking, the algorithm only stops after 67 iterations / 298 seconds. In general this can not only be applied to small objects of course. Most of the scans are centered in the middle of the domain, not protruding out of the unit ball (radius = width/2), so a masking with k = width/2 can almost always be applied.

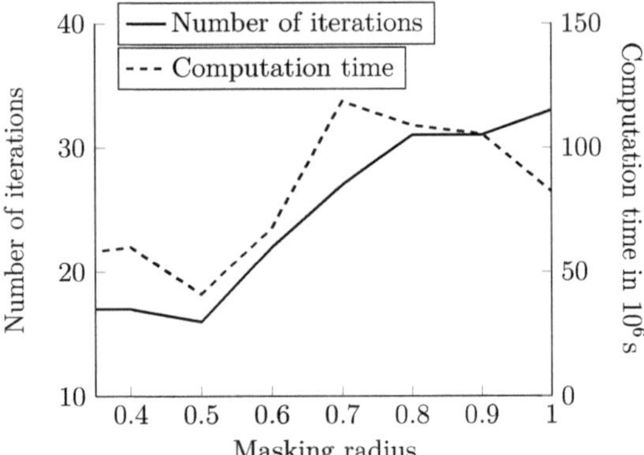

Figure 15: Number of iterations and computation time over varying masking radius.

8. Real 3D images

Real 3D images most of the time consist of a lot more gray value levels than the synthetic images that have been shown in this work so far. They might include noise and image artifacts. The algorithm has been tested on several input volumes taken from "The Volume Library" [Röt06] that offers volume datasets in PVM format. The datasets contain volume data acquired by MRI and CT and can be converted into a set of PGMs by the V^3 [Röt12].

First the application to the input volume "Stanford Bunny" will be discussed. The volume is a CT scan of the terracotta "Stanford Bunny" scanned by Terry S. Yoo, High Performance Computing and Communications, National Library of Medicine, USA. The intensity values denote the electron-density of the subject. The volume consists of 361 slices of 512x512 px.

When not limiting the amount of DOFs after 24 steps we have 5.382.066 DOFs with another 30.077.177 elements marked for refinement and only 12 elements marked for coarsening. On the attempt to adapt the mesh now, the algorithm crashes as a bus error occurs. Too much memory is requested to do the necessary calculations. Hence the amount of unknowns needs to be limited as mentioned in 3.3. When setting $maxDOFs$ to 200.000 the amount of DOFs will stay at 219.860 after 15 steps. The bunny is successfully segmented, the boundaries are detected correctly and holes that are located at the bottom of the bunny are filled. The result can be seen in Figure 16.

The data set includes some artifacts that are detected as boundaries and therefor segmented. To get better results the artifacts have been (partly) removed by hand, see Figure 17 and 18.

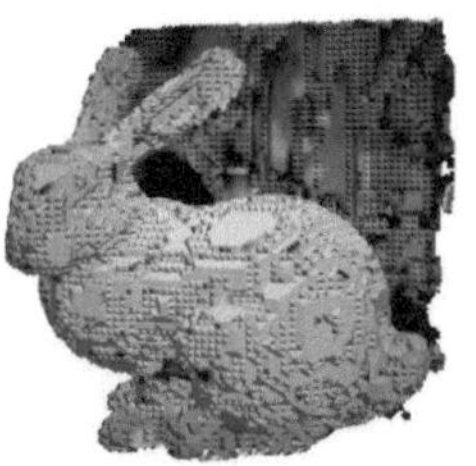

$\delta = 0.01$, 44 iterations, 10 min $\delta = 0.001$, 279 iterations, 47 min $\delta = 0.0001$, 633 iterations, 1 h 41 min

Figure 16: The segmented bunny with artifacts at different precision stages under a 0.01 threshold.

$\delta = 0.01$, 43 iterations, 6 min $\delta = 0.001$, 271 iterations, 45 min $\delta = 0.0001$, 400 iterations, 1 h 6 min

Figure 17: The segmented bunny at different precision stages under a 0.01 threshold. Most of the artifacts were removed manually prior to computation.

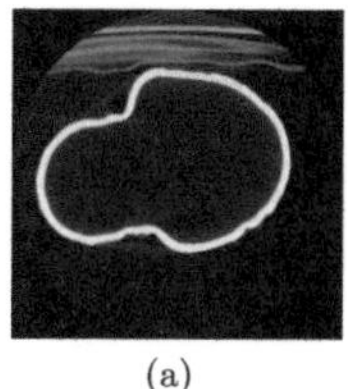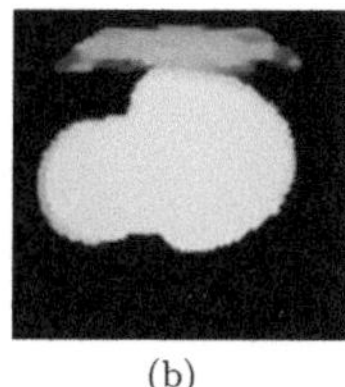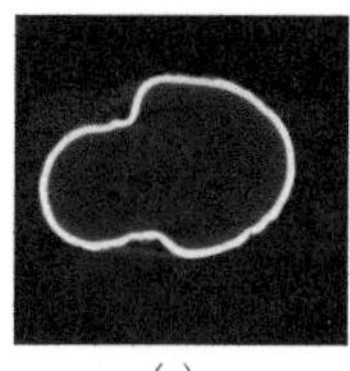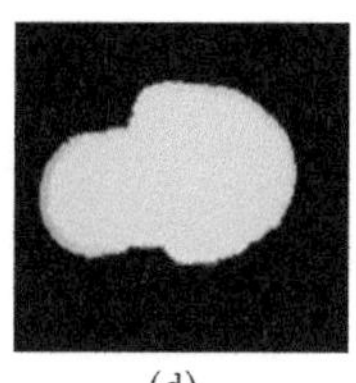

(a)   (b)   (c)   (d)

Figure 18: Slice 246 of (a) the original input volume, (b) the segmented bunny, (c) the input volume after partial artifact removal and (d) the corresponding segmentation

The same way CT scans of living things can be segmented. The algorithm has been applied to the volume "Baby Head" by Jason Bryan that is distributed with the VolSuite package and "Head (Visible Male)" from the The Visible Human Project, both found in the Volume Library. The outer boundary is correctly detected as to be seen in Figure 19 and Figure 20.

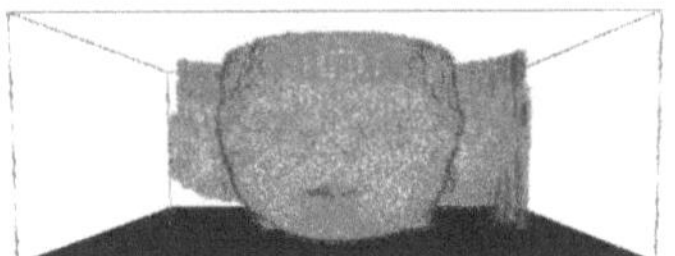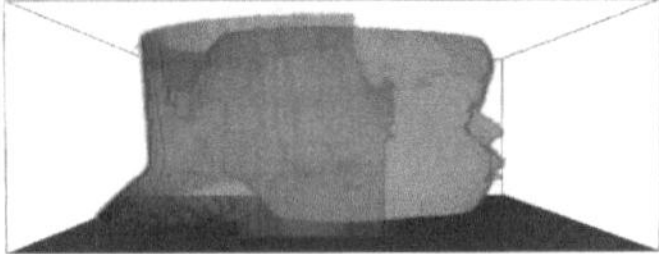

Figure 19: Segmentation of the volume "Baby Head".

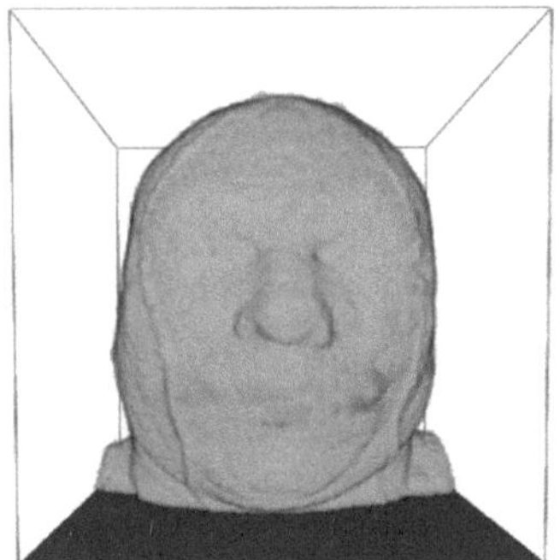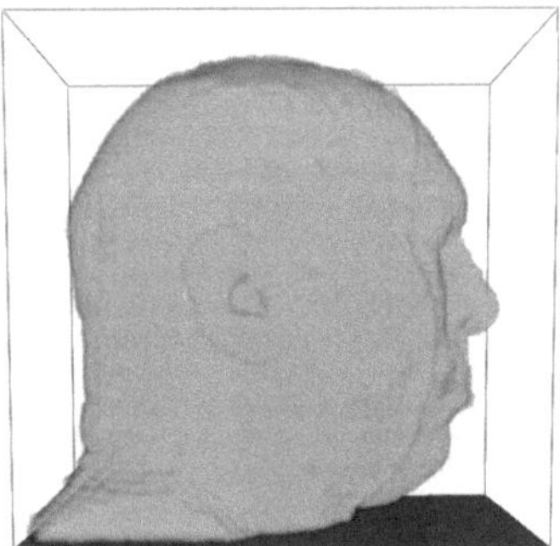

Figure 20: Segmentation of the volume "VisMale".

Part IV.
Conclusion and outlook

The aim of this work was to extend the 2D segmentation algorithm by Fried [FM09] to three dimensions and test whether it holds the same functions as in 2D, that is the ability to find real and modal boundaries, fill in holes in interrupted edges and work despite of noise or blurriness of edges. The algorithm relies on the subjective surface method solved with finite elements and adaptive mesh refinement. The main attention was directed to the 3D output, and the testing of the algorithm's capabilities. For that reason the algorithm was run on several synthetic images created by the author as well as on real 3D images taken from the Volume Library [Röt06].

With the developed tool it is possible to read in 3D volumes in the form of 2D slices in PGM format, segment the given objects and output them in different formats to inspect them either in ParaView, in V^3 or as 2D slices in PGM format.

In the tests presented in this work it became apparent that the algorithm is able to detect boundaries in synthetic and real 3D images with both noise and blurred regions. It also showed that missing boundaries are found and the modal completion is performed correctly.

A problem that arose is that for the 3D volumes, especially images with big gradients, the algorithm needs a lot of memory space as there are a many unknowns. Thus, in many cases the memory of one machine is not sufficient. Furthermore, calculating with a great amount of unknowns and setting the stopping tolerance very low to obtain good results leads to long computation times. Rewriting the algorithm to run parallel on a distributed memory system would speed up the computation tremendously and lead to more accurate results as the mesh could be further refined.

References

[BF12]   BOAS, F E. ; FLEISCHMANN, Dominik: CT artifacts: causes and reduction techniques. In: *Imaging in Medicine* 4 (2012), Nr. 2, S. 229–240

[BK04]   BARRETT, Julia F. ; KEAT, Nicholas: Artifacts in CT: Recognition and Avoidance 1. In: *Radiographics* 24 (2004), Nr. 6, S. 1679–1691

[CCCD93]   CASELLES, Vicent ; CATTÉ, Francine ; COLL, Tomeu ; DIBOS, Françoise: A geometric model for active contours in image processing. In: *Numerische mathematik* 66 (1993), Nr. 1, S. 1–31

[CKS97]   CASELLES, Vicent ; KIMMEL, Ron ; SAPIRO, Guillermo: Geodesic active contours. In: *International journal of computer vision* 22 (1997), Nr. 1, S. 61–79

[DD01]   DECKELNICK, Klaus ; DZIUK, Gerhard: Convergence of numerical schemes for the approximation of level set solutions to mean curvature flow. In: *SERIES ON ADVANCES IN MATHEMATICS FOR APPLIED SCIENCES* 59 (2001), S. 77–94

[DD03]   DECKELNICK, Klaus ; DZIUK, Gerhard: *Numerical approximation of mean curvature flow of graphs and level sets*. Springer, 2003

[ES+91]   EVANS, Lawrence C. ; SPRUCK, Joel u. a.: Motion of level sets by mean curvature I. In: *J. Diff. Geom* 33 (1991), Nr. 3, S. 635–681

[FM09]   FRIED, Michael ; MIKULA, Karol: *Efficient Subjective Surfaces Segmentation by Adaptive Finite Elements*. 2009. – Talk given at the "IMI International Workshop on Computational Photography and Aesthetics", Dec 13, Nanyang Technological University, Singapore

[Fri03]   FRIED, Michael: *Image segmentation using adaptive finite elements*. Math. Inst., 2003

[Fri04]   FRIED, Michael: A level set based finite element algorithm for the simulation of dendritic growth. In: *Computing and Visualization in Science* 7 (2004), Nr. 2, S. 97–110

[MS07]   MIKULA, Karol ; SARTI, Alessandro: Parallel co-volume subjective surface method for 3D medical image segmentation. In: *Deformable Models*. Springer, 2007, S. 123–160

[MSV95]   MALLADI, Ravi ; SETHIAN, James A. ; VEMURI, Baba C.: Shape modeling with front propagation: A level set approach. In: *Pattern Analysis and Machine Intelligence, IEEE Transactions on* 17 (1995), Nr. 2, S. 158–175

[OF01]   OSHER, Stanley ; FEDKIW, Ronald P.: Level set methods: an overview and some recent results. In: *Journal of Computational physics* 169 (2001), Nr. 2, S. 463–502

[Röt06]   RÖTTGER, Stefan: *The Volume Library*. http://www9.informatik.uni-erlangen.de/External/vollib/. Version: 2006. – [Online; accessed 10-February-2015]

[Röt12]   RÖTTGER, Stefan: V^3. http://www.stereofx.org/volume.html. Version: 2012. – [Online; accessed 10-February-2015]

[Set99]   SETHIAN, James A.: *Level set methods and fast marching methods: evolving interfaces in computational geometry, fluid mechanics, computer vision, and materials science*. Bd. 3. Cambridge university press, 1999

[SMS00]   SARTI, Alessandro ; MALLADI, Ravi ; SETHIAN, James A.: Subjective surfaces: A method for completing missing boundaries. In: *Proceedings of the National Academy of Sciences* 97 (2000), Nr. 12, S. 6258–6263

[SMS02]   SARTI, Alessandro ; MALLADI, Ravi ; SETHIAN, James A.: Subjective surfaces: a geometric model for boundary completion. In: *International Journal of Computer Vision* 46 (2002), Nr. 3, S. 201–221

[SS05]   SCHMIDT, Alfred ; SIEBERT, Kunibert G.: *Design of adaptive finite element software: The finite element toolbox ALBERTA*. Bd. 42. Springer, 2005

Parameter file

```
% Macro triangulation
macro file name:         Macro/macro_big.stand neum
global refinements: 4
polynomial degree:  1
lower left 0:        -1. %offset
lower left 1:        -1.
lower left 2:        -1.

% Solution of LSE
solver:                  2 % 1: BICGSTAB 2: CG 3: GMRES 4: ODIR 5: ORES
solver max iteration: 1000
solver tolerance:        1.e-4
solver info:             2
solver precon:           1 % 0: no precon 1: diag precon

% Error estimation
adapt->tolerance:          0.125
adapt->timestep:           0.005
adapt->rel_initial_error:  1.
adapt->rel_space_error:    1.
adapt->rel_time_error:     2.
adapt->max_iteration:      1
adapt->info:               3
estimator C0:              0.0
estimator C1:              1.0
estimator C2:              1.0
estimator C3:              1.0

% Initial mesh adaption
adapt->initial->strategy:                3 % 0=none, 1=GR, 2=MS, 3=ES, 4=GERS
adapt->initial->max_iteration:           10
adapt->initial->info:                    10
adapt->initial->coarsen_allowed:         1
adapt->initial->refine_bisections:       2
adapt->initial->coarsen_bisections:      1

% Mesh adaption
adapt->space->strategy:          4       % 0=none, 1=GR, 2=MS, 3=ES, 4=GERS
adapt->space->GERS_theta_star:  0.75
adapt->space->GERS_nu:          0.1
adapt->space->GERS_theta_c:     0.25
adapt->space->max_iteration:    1
adapt->space->coarsen_allowed:  1

% Time discretization
```

```
theta:                 1.0 % 1=explicit , 0.5 =semi-implicit , 0=implicit
adapt->end_time: 10.0

% Evans-Spruck regularization
epsilon:               1.0
epsilon scaling: 30

% Stopping criterion
delta stop: 0.001

% Smoothing steps
smooth: 0

% Input
pgm file: PGMS/TestVolume/test
number of pgms: 100

% Output
pgm outputfile: testSegmentation
```

YOUR KNOWLEDGE HAS VALUE

- We will publish your bachelor's and
  master's thesis, essays and papers

- Your own eBook and book -
  sold worldwide in all relevant shops

- Earn money with each sale

Upload your text at www.GRIN.com
and publish for free